Sweety Tweets
IN A Fast-Paced World

How to Stay Connected in
120 Characters or Less!

Kathleen Barbier

ISBN:1463738978
ISBN-13: 9781463738976

Sweety Tweets in a Fast-Paced World

How to Stay Connected in 120 Characters or Less!

Kathleen Barbier

✒ *Table of Contents*

Introduction ✑

Today, we are *RACING* through life at warp speed, often **OVERWHELMED** and **Stressed Out** with too much to say, do, and think, let alone communicate.

There are too many texts, tweets, phone calls, billboards, TV Ads, and other communication "noise" assaulting us 24/7 in our light-speed world.

We no longer have time for long emails or phone calls! After all, we often only have 5-10 seconds to communicate, which translates electronically into 120 characters or less.

We stop communicating with our loved ones. But, is that really what we want?

What we will NEVER HAVE ENOUGH OF is short, loving "just wanted to let you know I was thinking of you" messages from those who are most important to us.

We need to keep our relationships fresh and alive, and to STAY CONNECTED.

Let me share with you a Contemporary Fairy Tale:

Contemporary Fairy tale

A Twenty-First Century Fairy Tale Gone Bad, Then Saved --

Once Upon a Time, in a faraway place, Prince Robert met Princess Kathleen on an electronic dating site.

Everything went perfectly with this Match (hint) until they tried to fit their new-found relationship into their everyday living.

ALAS, to their mutual consternation, the communication started to break down as Prince Robert TRIED to communicate his undying love via email.

Princess Kathleen, being the contemporary Princess that she was/is, began to see long (we could define this more precisely, but I think you get the picture, as I see that you are already chuckling) emails on her Smart Phone from the moment she opened her wee Princess eyes in the morning.

DOUBLE ALAS, as even before sipping her morning Mocha, Princess Kathleen's waking brain began to short circuit. Sizzle. Sputter. ZZzzzz. Crash!

GULP. The messages came often, with great, loving detail, resulting in a communication overload for Princess Kathleen, who already received 200+ emails, texts, calls, and Faxes in her Fast-Paced World.

Her love for Prince Robert was being clouded.

Her knee-jerk reaction (not Princess-like) was to send a counter reply … "Please, I don't need to know your every waking thought."

Yes, it is sad, but true that the hastily sent email struck her Prince's loving heart squarely in its hopeful, gently beating core, striking a blow against true love.

Alas, even a Princess has her "bad hair" days!

REACTION, NOT RESPONSE was the order of the day as Princess Kathleen first declared a moratorium on all emails coming from her loving Prince Robert.

Ah. What to do?   Prince Robert was sad that he could not express his loving thoughts to his fair Princess.

In this Tale, it was Princess Kathleen who had turned into the Toad. RIB-itt. RIB-itt. (A toad Sweety Tweet)

The Solution – A Game is Born

BUT, YET, the Sun again arises in the Kingdom of Happiness and Joy as a new game is born.

The repentant Princess Kathleen made a decree upon the land that, from here forward, she and Prince Robert would have a game to play to keep their loving communications alive and well.

This game would be their own private game, and it would enable them to stay connected in a high-speed, cyber-paced world.

THE NEW GAME WAS CALLED "SWEETY TWEETS", with guidelines similar to those we routinely use in our daily lives.

The Rules

The game rules were easy. All Sweety Tweets would have no more than 120 characters, but each Sweety Tweet sender gets unlimited rights to send as many as he or she wants throughout the day.

Bugles sounded throughout the land.

Can we see relationships flourishing, and Best Friends Forever (BFFs) responding with joy and laughter?!

AND, AS THE SUN SETS IN THE ELECTRONIC DATING KINGDOM where our fair Princess and her Prince continue their match unabated, we see Prince Robert and Princess Kathleen happily exchanging snippets of information whenever their loved one comes to mind.

The sun again rises.

Now, doesn't this remind you of being younger and passing a note in class without getting caught?

Imagine a Sweety Tweet sneaking into the middle of your classroom or office! Such joy.

TODAY, THE PRINCE AND HIS PRINCESS HAPPILY open their Sweety Tweets the moment they receive them, knowing that it will be a short, entertaining, loving exchange.

It will provide a "Power Boost" for the day. It will make them smile.

The recipient knows that he or she is loved, or even a valued BFF and life is good.

In a world that discounts us all, the Sweety Tweet receivers KNOW their true worth.

Princess Kathleen has even declared it a better way to start her day than her traditional Mocha!!

HARMONY ONCE AGAIN CAME OVER THEIR KINGDOM.

And, Prince Robert and Princess Kathleen live happily ever after, sending 2.5 Sweety Tweets per day, on average, at a new "EXPERT SKILL LEVEL" of 45 characters or less!

We see a rainbow across the sky as Prince Robert leans forward to kiss his blushing Princess.

The End

Purpose

The purpose of this book is to inspire loving interchanges in relationships constrained by time, whether those relationships are brand new or long enough to have attained BFF status and beyond.

We understand that electronic dating sites are giving the book freely to their matches, hoping that the new, enhanced communication will increase their client satisfaction ratings among their matches.

Sweety Tweets have energized our connections, and replaced texts, emails and even phone calls as the preferred couple talk tool of the twenty-first century!

And, in the words of Princess Kathleen …

Let the Sweety Tweets begin!

May the Force be with you, and your Sweety Tweets keep peace in your Kingdom!

Princess Kathleen

Sweety Tweets for the Novice

Getting Started

Directions:

The purpose of this book is to get you started in the Sweety Tweets game with minimal start up time.

Therefore, this book is filled with a "starter set" of Sweety Tweets, which are 120 characters or less in length.

Where do you start?

If you're not quite sure how to start, look through the Table of Contents, and find the category that speaks to you.

Follow your heart, and you will know where to begin.

Before long, you will be Sweety Tweeting with the best of them!

Also included are some "Expert Level" examples, using 45 characters or less, when time is of the essence!

STEP 1 – SEND THE FIRST SWEETY TWEET

To get started: Type "Sweety Tweets Program" in the Subject Line.

Subject: Sweety Tweets Program:

(Note: Text for the program explanation would follow)

Hi, _____ (name), I'm sending a Sweety Tweet to brighten your day! It'll always be short (<120 characters), so enjoy!

Characters: 117

STEP 2 – IMMEDIATELY FOLLOW WITH ANOTHER SWEETY TWEET.

Hint: Lead by example.

Situation: In this example, conflict is growing over a misunderstanding as one person sends long emails, but the other doesn't have the time to fully read and comprehend them, let alone respond to them!

Sweety Tweet Example (Conflict)

Subject: Sweety Tweet for You

Don't be so hard on yourself. We're just different, and I like that about us!

Characters: 78

101 Sample Sweety Tweets

The following pages are filled with examples of Sweety Tweets that can be used as is, modified, or used to inspire.

Be creative. Be sincere. But, most importantly, be loving and succinct!

Remember, the 120 characters or less you choose to send will shape your emotional well-being today, tomorrow, and in the future.

The two minutes you take to send a Sweety Tweet will add "money in the bank" with your sweetheart.

It is the wisest investment of your time that you can possibly make.

When you consider that relationships can thrive for 30 or 40 years and that a BFF is, by definition, forever, imagine what is possible!

A Sweety Tweet of apology will be worth its weight in gold.

A flirty Sweety Tweet sets the stage for a glorious evening.

It is all up to you.

What will it be?

To Deal with Conflict

Sweety Tweet #1 (Early Morning Conflict)

Sorry we started the day with a misunderstanding. Forgive me?

Characters: 62

℘

Sweety Tweet #2 (Conflict)

Whatever I did this morning to get us started on the wrong foot isn't as important as YOU are to me. I apologize!

Characters: 114

℘

Sweety Tweet #3 (Verbal Conflict)

Something really stupid fell out of my mouth this morning, but I can't remember what it was! Hoping you can't either!

Characters: 118

℘

Sweety Tweet #4 (Conflict Recovery)

*They say chocolate heals all.   Will it take a 1#
or a 2# box to recover from this morning?*

Characters: 91

❧

Sweety Tweet #5 (Conflict Recovery)

*Hard to believe someone so bald could have
such a "bad hair" day. Forgive me?*

Characters: 78

❧

Sweety Tweet #6 (Conflict Recovery)

*If I give up the TV clicker for a week, will you
forgive me?*

Characters: 60

❧

Sweety Tweet #7 (Conflict)
EXPERT LEVEL

Sorry. Can you forgive me?

Characters: 27

Sweety Tweet #8 (Conflict Recovery)
EXPERT LEVEL

Dinner out, or dog house in, tonight?

Characters: 37

Sweety Tweet #9 (Conflict Recovery)
EXPERT LEVEL

Flowers, movie, or grovel?

Characters: 26

Sweety Tweet #10 (Conflict Recovery)
EXPERT LEVEL

Sorry beyond words. Let me count the ways!

Characters: 42

To Start the Day off Right

Sweety Tweet #11 (Morning Thought)

Just wanted to let you know I am thinking about you this morning!

Characters: 65

♉

Sweety Tweet #12 (Morning Break)

It made me smile as I thought about you this morning.

Characters: 53

♉

Sweety Tweet #13 (Thoughts of You)

This morning is filled with thoughts of seeing you tonight! Can't wait.

Characters: 72

♉

Sweety Tweet #14 (Morning Leads to Lunch)

The best thing about this morning is that it fills the time until I have lunch with you!

Characters: 88

℘

Sweety Tweet #15 (Morning Break) EXPERT LEVEL

Thinking of you as my morning break.

Characters: 36

℘

Sweety Tweet #16 (Morning Thought) EXPERT LEVEL

Just thinking of you jumpstarts my day!

Characters: 39

℘

Sweety Tweet #17 (Rising Sun)
EXPERT LEVEL

The rising sun brought thoughts of you.

Characters: 39

Sweety Tweet #18 (Morning Start)
EXPERT LEVEL

Thoughts of you are better than Mocha!

Characters: 38

Sweety Tweet #19 (Thoughts of You)
EXPERT LEVEL

Had to catch my breath as I thought of you.

Characters: 43

Important "Morning After" Sweety Tweet

Sweety Tweet #20 (Morning After)

Last night added new meaning to "seeing stars"! Can't wait to see you again.

Characters: 77

Sweety Tweet #21 (Morning After)

Just had unscheduled surgery to remove smile from face and reconnect brain.

Characters: 75

Sweety Tweet #22 (Morning After – Blue Eyed)

This morning I feel blue like the color of your eyes without you.

Characters: 65

Sweety Tweet #23 (Morning After – New Year's)

What a way to start the New Year!

Characters: 33

❦

Sweety Tweet #24 (Morning After – Christmas)

I must have made it to Santa's "Nice" list this year!

Characters: 53

❦

Sweety Tweet #25 (Morning After – 4th of July)

The best fireworks last night were those set off by you.

Characters: 56

❦

Sweety Tweet #26 (Morning After)

Thanks for a spectacular night. Been thinking of you all day.

Characters: 62

Q

Sweety Tweet #27 (Morning After)

Think we may have qualified for Ripley's Believe it or Not last night?

Characters: 70

Q

Sweety Tweet #28 (Morning After)

Had a great dream last night, or was that us?

Characters: 45

Q

Sweety Tweet #29 (Morning After)
EXPERT LEVEL

You've left me speechless.

Characters: 26

❦

Sweety Tweet #30 (Morning After)
EXPERT LEVEL

Bravo! Repeat performance soon??

Characters: 34

❦

Sweety Tweet #31 (Morning After)
EXPERT LEVEL

Unprecedented and unparalleled!!

Characters: 33

❦

Sweety Tweet #32 (Morning After) EXPERT LEVEL

Unbelievable!!

Characters: 15

❦

Sweety Tweet #33 (Morning After) EXPERT LEVEL

Your place or mine?

Characters: 19

❦

Sweety Tweet #34 (Morning After) EXPERT LEVEL

A new, personal best!

Characters: 21

❦

Sweety Tweet #35 (Morning After)

Last night was the magic that dreams are made of!

Characters: 49

Sweety Tweet #36 (Stunned)
EXPERT LEVEL

I'm rarely stunned and speechless.

Characters: 34

When You Have to Cancel

Sweety Tweet #37 (Cancel Breakfast)

I'm devastated that I won't be able to start the day with breakfast with you! Reschedule?

Characters: 90

❧

Sweety Tweet #38 (Cancel Breakfast)

Hate to start the day without seeing you, but was called in early. Forgive me?

Characters: 79

❧

Sweety Tweet #39 (Cancel Breakfast)

How can I miss the best meal of the day with my best girl? Reschedule?

Characters: 70

❧

Sweety Tweet #40 (Cancel Lunch)

An apple a day is not a substitute for lunch with YOU. Forgive me?

Characters: 67

Sweety Tweet #41 (Cancel Lunch)

Sorry I got caught in an unscheduled "pep talk" at work. Reschedule for Friday?

Characters: 81

Sweety Tweet #42 (Cancel Lunch)

Just got hit with major project that wiped out our lunch. Rather be with you. Forgive me?

Characters: 91

Sweety Tweet #43 (Cancel Lunch)

How am I supposed to function without seeing you at lunch? Called into meeting. Dinner tonight?

Characters: 97

Sweety Tweet #44 (Cancel Dinner)

Darn! Pulled into meeting. Have to skip dinner with best girl? I'll call when I escape!

Characters: 90

Sweety Tweet #45 (Cancel Multiple Dinners)

If I miss one more dinner with you, I'm going to owe you a great 3-day weekend!

Characters: 79

Sweety Tweet #46 (Cancel Dinner & Reschedule)
EXPERT LEVEL

Sorry. Reschedule to favorite restaurant?

Characters: 42

Sweety Tweet #47 (Cancel Dinner & Paris)
EXPERT LEVEL

Reschedule to Paris, s'il vous plait?

Characters: 37

Sweety Tweet #48 (Cancel Drinks After Work)
EXPERT LEVEL

Drew the short straw at work! Next week?

Characters: 41

Sweety Tweet #49 (Cancel Drinks) EXPERT LEVEL

Catching early plane tomorrow. Next week OK?

Characters: 45

Sweety Tweet #50 (Cancel Coffee) EXPERT LEVEL

Have to backstop someone at work. Next week?

Characters: 45

Romantic Sweety Tweets

Sweety Tweet #51 (Romantic Rendez-vous)

Have appointment near you today. Seeing you mid-day would rock. Time for a quick rendezvous?

Characters: 94

℧

Sweety Tweet #52 (Romantic)

You've captured my heart, and I can't help but declare "I love YOU!"

Characters: 68

℧

Sweety Tweet #53 (Romantic)

The 3 most beautiful words I'll say in this life-time…I love you.

Characters: 64

℧

Sweety Tweet #54 (To a Blonde)

Repunzel, Repunzel, let down your golden hair. Must see you!

Characters: 61

Sweety Tweet #55 (Wishes)

Your love opened my heart so completely that all my saved-up wishes have just come true.

Characters: 89

Sweety Tweet #56 (Childhood Joy)

Whenever I think of you, the racing pitter-patter of my heart reminds me of dancing footsteps.

Characters: 94

Sweety Tweet #57 (Romantic Wait)

A thousand days and nights were not too long to wait for you. It only seemed forever!

Characters: 86

Sweety Tweet #58 (Love Waiting)

I've waited forever for you to steal my heart, and know you will return it unharmed!

Characters: 84

Sweety Tweet #59 (Love Songs)

As I look into your eyes, my heart begins singing love songs. I'll be ready for American Idol soon!

Characters: 100

Sweety Tweet #60 (Dream Come True)

My love grows stronger with every dream you've made come true!

Characters: 62

Sweety Tweet #61 (Snow White)

You needn't look into your mirror to know what I see, that you are the fairest maiden in the land!

Characters: 99

Sweety Tweet #62 (Sleepless in Seattle)

Like Sleepless in Seattle, I felt like I was coming home again. Looking forward to seeing you soon.

Characters: 100

Sweety Tweet #63 (Coins in Fountain)

I've dreamed for years of tossing 3 coins in a fountain and making a wish. They have now come true.

Characters: 100

Sweety Tweet #64 (Laugh of Joy)

Thinking of you today brought joyful laughter and a smile to my face.

Characters: 69

Sweety Tweet #65 (Singing)

My neighbors keep asking what makes me sing. Just know that it is YOU!

Characters: 71

Sweety Tweet #66 (Tongue Tied)

My appreciation of you is exceeded only by an inability to fully express it.

Characters: 76

❦

Sweety Tweet #67 (Snow White)
EXPERT LEVEL

Snow White, abandon the dwarfs; meet me at 8!

Characters: 45

❦

Sweety Tweet #68 (First Hello)
EXPERT LEVEL

You had me at the first "hello".

Characters: 32

❦

No Time to Call

Sweety Tweet #69 (No Time to Talk)

Wanted to call just to hear your voice, but my cell is dying. Kisses.

Characters: 70

Sweety Tweet #70 (No Time to Talk)

Would call to hear your sexy voice, but I'd be distracted all day!

Characters: 67

Sweety Tweet #71 (No Time to Talk)

Would call if I had even one free second, but will send this kiss instead!

Characters: 74

Sweety Tweet #72 (No Time to Talk)

The sound of your laughter still lingers in my mind. Call later.

Characters: 65

Just Because

Sweety Tweet #73 (Just Because)

Keep clicking heels and saying "there's no place like home." Must have wrong shoes on!

Characters: 87

Sweety Tweet #74 (Just Because)
EXPERT LEVEL

Wanted you to know you're on my mind!

Characters: 37

Sweety Tweet #75 (Just Because)
EXPERT LEVEL

Just thinking of you, Sweetheart!

Characters: 33

Sweety Tweet #76 (Just Because)
EXPERT LEVEL

Kisses to my one in a million!

Characters: 30

Sweety Tweet #77 (Just Because)
EXPERT LEVEL

Be still my heart!

Characters: 18

Time to Celebrate

Sweety Tweet #78 (Celebrate Salary Bonus)

Would trade bonus I just earned for 5 minutes with you. Meet me at our favorite place to celebrate?

Characters: 100

Sweety Tweet #79 (Surprise Celebration)

Too excited to spoil the surprise. Meet me at our favorite place!

Characters: 66

Sweety Tweet #80 (1 Month Anniversary)

Time to celebrate your coming into my life. Hard to believe it's only been a month!

Characters: 84

Sweety Tweet #81 (1 Year Celebration)

Hard to believe my great fortune as you took my breath away a year ago. Meet to celebrate?

Characters: 91

Ⓠ

Sweety Tweet #82 (6 Month Anniversary)

Six months of joy and happiness. Let's go away this weekend to celebrate!

Characters: 74

Ⓠ

Sweety Tweet #83 (1 Year Anniversary)

One year of complete joy. Let's steal away early from work today to celebrate, My Love!

Characters: 88

Ⓠ

Sweety Tweet #84 (2+ Year Anniversary)

Darling: Thank you for the best ___ years of my life. Can't wait to celebrate tonight!

Characters: 88

✐

Asking for Help

Sweety Tweet #85 (Pick up Laundry)

Would be lost without you. No chance to pick up laundry before I leave in the am. Can you save me?

Characters: 100

❦

Sweety Tweet #86 (Doggie Sitting)

Rover asked if you can visit while I'm gone. He promised to keep an eye on you for me!

Characters: 87

❦

Sweety Tweet #87 (Forgotten Kids' Weekend)

Forgot it was my weekend to play Dad. Can you join us for a day in the park?

Characters: 77

❦

When Feeling Rebellious

Sweety Tweet #88 (Rebellious)

Watch for an email today. Can't tell you how wonderful you are in 100 characters or less!!

Characters: 91

Sweety Tweet #89 (Feisty)

Know you said something really nice to me this morning, but think I heard it differently, Princess!

Characters: 99

Sweety Tweet #90 (Feisty)

Know you thanked me for all the weekend work I did, but I can't remember hearing it.

Characters: 84

Sweety Tweet #91 (Feisty & Memory Problem)

If I could remember what I was angry about, I would apologize. How about "Love you" instead?

Characters: 93

Checking the Temperature

Sweety Tweet #92 (Pre-call Check After a Fight)
EXPERT LEVEL

Having a good day?

Characters: 18

Sweety Tweet #93 (Home Late Pre-Call Check)
EXPERT LEVEL

Has that beautiful smile returned?

Characters: 34

Sweety Tweet #94 (Making Up Roses)
EXPERT LEVEL

Trade red roses for badly-needed hug?

Characters: 37

Sweety Tweet #95 (Something Stupid)
EXPERT LEVEL

Was that a nightmare, or me acting out?

Characters: 39

Sweety Tweet #96 (Forgiveness)
EXPERT LEVEL

Pleeeeeeeease forgive me (again)?!

Characters: 35

When You're Running Late

Sweety Tweet #97 (Running Late)
EXPERT LEVEL

Runny late, Fair Maiden; please don't escape!

Characters: 45

Sweety Tweet #98 (Late)
EXPERT LEVEL

I'm late. I'm late for a very important date!

Characters: 45

Sweety Tweet #99 (Late)
EXPERT LEVEL

Sorry. No words can express… Be there soon!

Characters: 44

Sweety Tweet #100 (Late)
EXPERT LEVEL

Princes are NEVER to be late, M'Lady. Sorry!

Characters: 45

❧

Sweety Tweet #101 (Late)
EXPERT LEVEL

No excuses for making you wait, just regrets.

Characters: 45

❧

SUMMARY & CLOSING REMARKS:

It is important to pay attention to the relationship between the seriousness of the issue at hand, and the brevity of the Sweety Tweet.

The more serious the issue, the more important it is to **keep the message short!**

Too many words increase the odds of a misunderstanding when someone is overly stressed.

Good Luck and Good love!

Princess Kathleen

KATHLEEN BARBIER parlayed a successful business career as an executive with Fiserv, Inc. responsible for Marketing, Sales, Client Services, and Corporate Communication, into the establishment of Estate Doctor®, a wealth management business focused on helping women who have undergone sudden loss, and who must for the first time in their lives learn how to manage their financial affairs.

Born in Carmel, California, Kathleen earned advanced business degrees (MBA and a Master of Science – Finance), taught graduate and undergraduate School of Business courses, and became a skilled change agent in the public, private, and non-profit sectors.

Kathleen believes that communication is the key to much of life's challenges. While working in the public sector, Kathleen had the opportunity to work with hundreds of city employees to help them embrace change and let go of the things holding them back.

Kathleen lives in Pebble Beach, California and has two grown children who are married.